THUNDEROUS STORMS

Samuel McNabb

TABLE OF CONTENTS

INTRODUCTION

Thunderstorms are essential to life here on Earth because they are a major source of rainfall. Although these storms can cause lots of damage, they provide much of the fresh water that plants need. We need thunderstorms, but we also need to know as much as we can about them so we can keep ourselves safe.

In the folklore of many cultures, thunder is said to come from the gods. Some people who practice a Japanese religion called Shinto say thunder comes from a spirit beating on a drum. The Vikings believed that thunder and lightning were caused by the god Thor slamming down his hammer.

Thunderstorms can be hard to predict without the right tools. Scientists have developed technological innovations like weather satellites and Doppler radar to increase how reliably they can predict these powerful storms. Many countries have dedicated huge amounts of money to agencies that help create reliable forecasts and weather models. Some examples are the National Weather Service in the United States and the European Weather Service that serves most countries in the European Union.

With all the information these groups and their new technology can provide, how much do we know about thunderstorms?

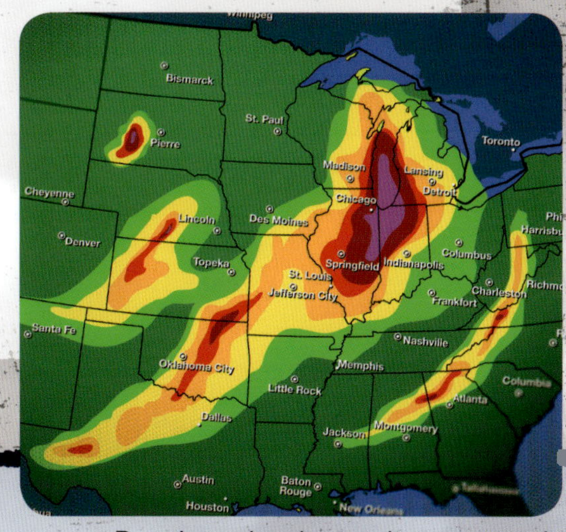

Doppler radar shows where a storm is and where it is going.

Scientists who study thunderstorms send up weather balloons that collect data on how lightning forms so they can learn how to predict how severe a storm will get. They also use tools that measure wind speed and direction close to the ground to learn how to predict tornadoes.

WHAT CAUSES THUNDERSTORMS?

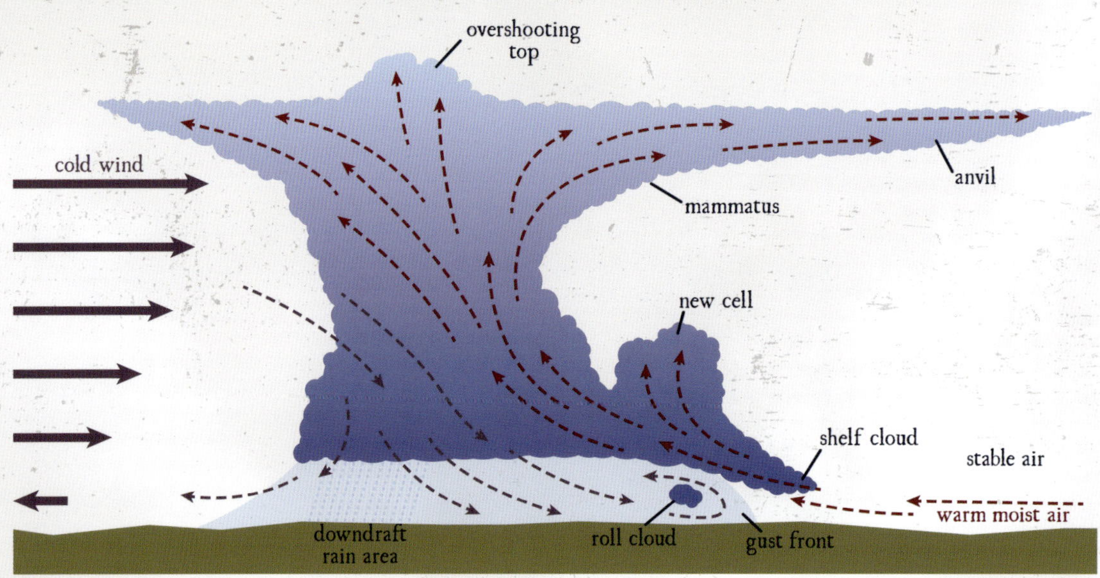

One ingredient of thunderstorms is warm, moist air moving upward through the atmosphere. The vertical lifting of the air can be caused by three things:

- the uneven warming of the Earth's surface,
- air flowing over a large range of mountains, or
- two air masses colliding at what's called a frontal zone.

Because there is so much moisture in the rising air, big clouds form as the water vapor cools and condenses. The clouds get darker as more and more droplets make them heavy.

Once it starts to rain, what makes a storm a thunderstorm, rather than just a rain shower? There needs to be cool, dry air flowing downward from the cloud toward the Earth. The combination of the warm, moist air pushing upward and the cool, dry air flowing downward creates an electric charge that causes lightning. During a lightning strike, the air gets hot suddenly, and it expands rapidly. That expansion makes the rumble or boom that we call thunder.

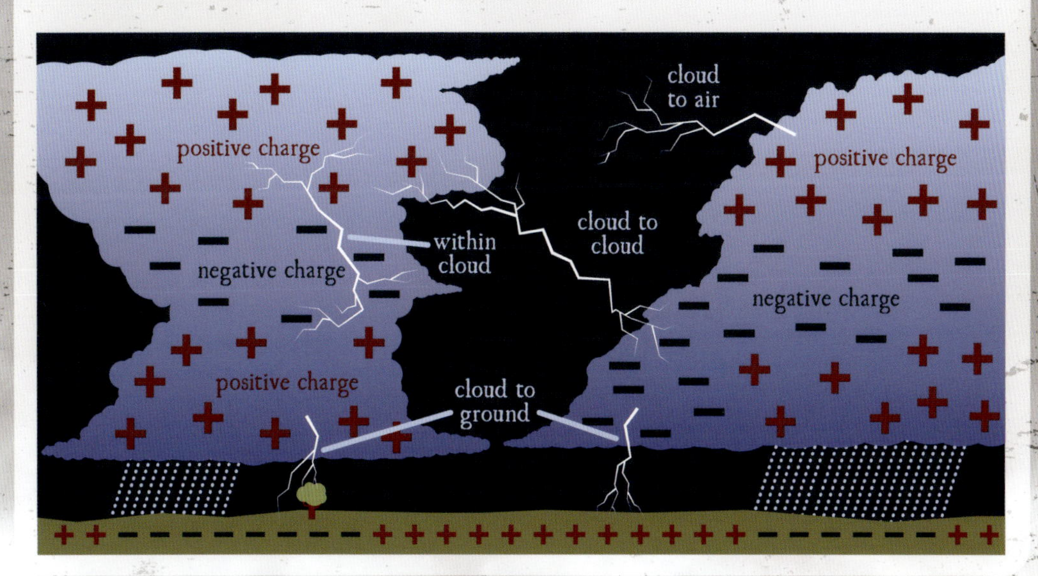

Unstable air can happen all over the world, so thunderstorms can, too. Thunderstorms can even form in one part of the world and travel somewhere else in a jet stream, an air current that circles the Earth.

Polar Jet Stream
Subtropical Jet Stream

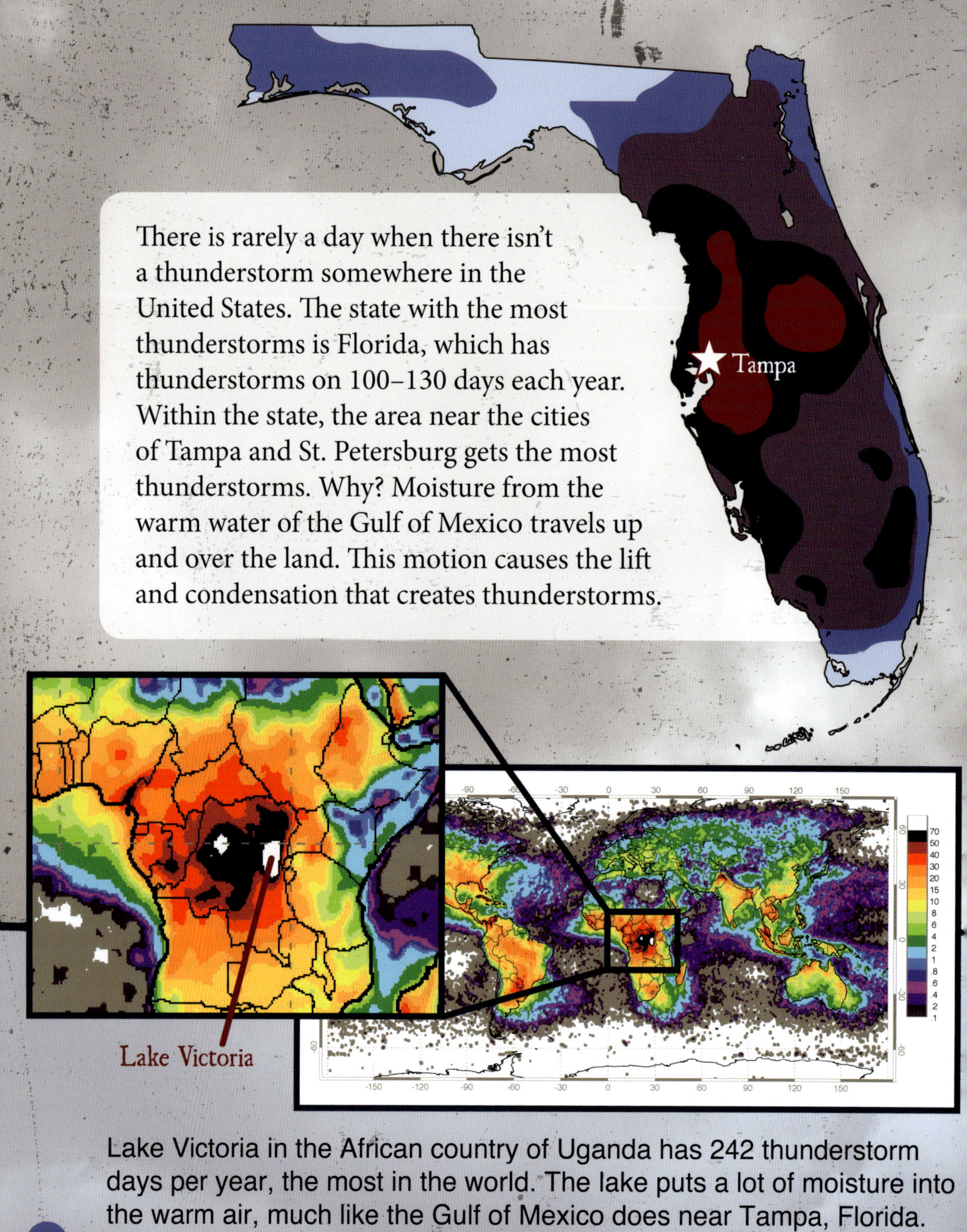

There is rarely a day when there isn't a thunderstorm somewhere in the United States. The state with the most thunderstorms is Florida, which has thunderstorms on 100–130 days each year. Within the state, the area near the cities of Tampa and St. Petersburg gets the most thunderstorms. Why? Moisture from the warm water of the Gulf of Mexico travels up and over the land. This motion causes the lift and condensation that creates thunderstorms.

Lake Victoria in the African country of Uganda has 242 thunderstorm days per year, the most in the world. The lake puts a lot of moisture into the warm air, much like the Gulf of Mexico does near Tampa, Florida. When the moist air is stirred by winds that blow from many directions toward the center of the lake, the conditions are right for thunderstorms.

THUNDERSTORMS ARE PART OF THE WATER CYCLE

The amount of water on Earth never changes. It just changes form in an endless process called the water cycle.

Each day, about four trillion gallons of water fall to Earth as rain, snow, sleet, or hail. Some of the water soaks into the ground, and some collects in rivers, lakes, and oceans. The rest—more than 2.5 trillion gallons—returns to the atmosphere through evaporation. As the water vapor rises, it cools and condenses. The water droplets join to form clouds. When a cloud becomes too heavy to hold on to its water droplets, the droplets fall to the ground as a form of precipitation.

WATER CYCLE

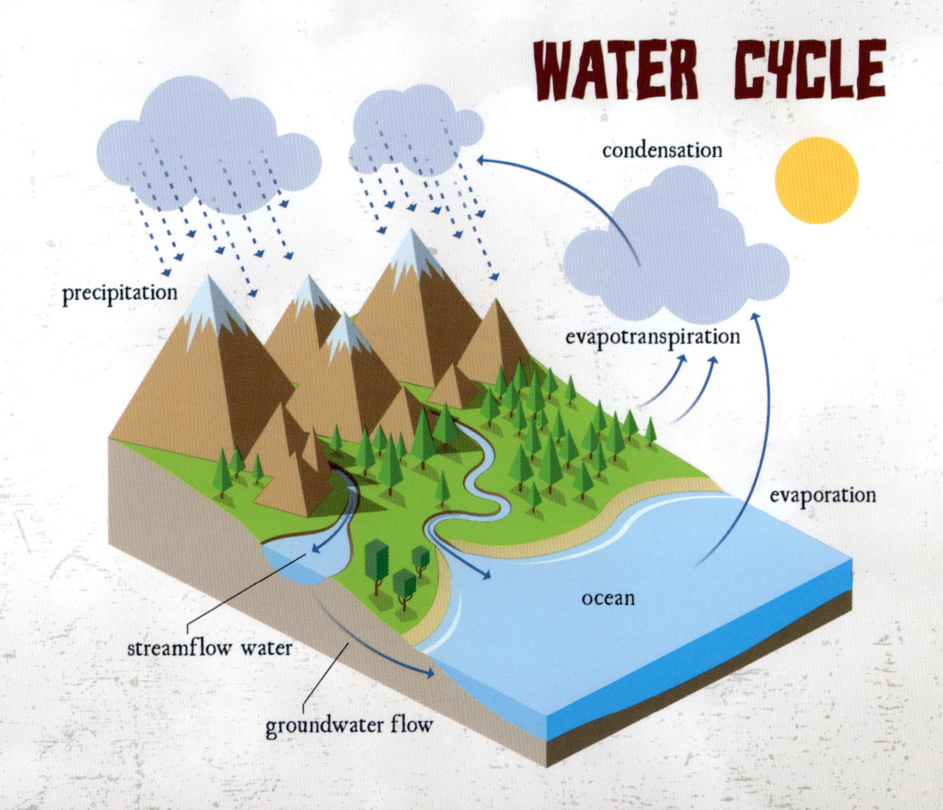

Most of the water on the planet is salt water, which is not very useful for plants, people, and most other animals. Because salt does not evaporate with the water, the water cycle turns huge amounts of salt water from the oceans into fresh water. Some of that fresh water falls during thunderstorms. That makes thunderstorms an important part of the water cycle.

When snow falls during a thunderstorm, it is called thundersnow.

THUNDERSTORMS ARE DANGEROUS

Thunderstorms are useful sources of fresh water, but they also cause plenty of problems. The lightning, wind, and rain from a thunderstorm can cause dangerous conditions for people, animals, and property.

Lightning from a thunderstorm can cause fires, including forest fires. Forest fires may start small, but some grow to cover many acres of land. They can ruin forests and destroy the homes of people and animals. Sometimes, forest fires are deadly. A forest fire in Oakland, California, killed more than 25 people. It also destroyed 2,843 houses and 433 apartments.

A sensor on the International Space Station records information about lightning strikes, including the timing, location, and amount of energy released. An average lightning bolt contains enough energy to power a 60-watt lightbulb for six months.

The wind from thunderstorms can be dangerous, too. During a severe thunderstorm, the wind can knock down trees and rip roofing and siding off buildings. Some thunderstorms cause very large windstorms called derechos that have winds up to 100 miles per hour. Derechos usually spread over a very wide area, which makes them incredibly destructive. Thunderstorms can even spawn tornadoes. The strong, rotating winds of a tornado can destroy a whole town, causing millions of dollars in damage and the loss of life.

In desert regions, the strong winds from a thunderstorm can cause a dust storm called a haboob. Haboobs can carry large debris, and they can travel a long way. During the 1930s, one haboob that started in the midwestern United States traveled all the way to New York City.

As dangerous as lightning and wind can be, rain causes the most deaths during thunderstorms. Heavy rains can cause flash floods that destroy entire communities and take many lives. Weather forecasters work hard to predict flash flooding so they can issue flash flood watches and warnings. These alerts can help keep people and communities safe if they pay attention to them.

In weather forecasting, a watch means that a bad weather event is possible in the area. A warning means that the weather event is already happening.

THUNDERSTORMS CAN HARM BUSINESSES

Thunderstorms don't just damage property. They can harm businesses, too. Companies that provide transportation often feel the negative effects of a thunderstorm. Cargo and passenger trains can be delayed because of fallen trees or power lines. Trucks that move cargo across the country may have to get off the road until a storm passes. Ships and planes may have to change their travel plans to avoid a thunderstorm. Delays like these cost businesses a lot of extra time and money.

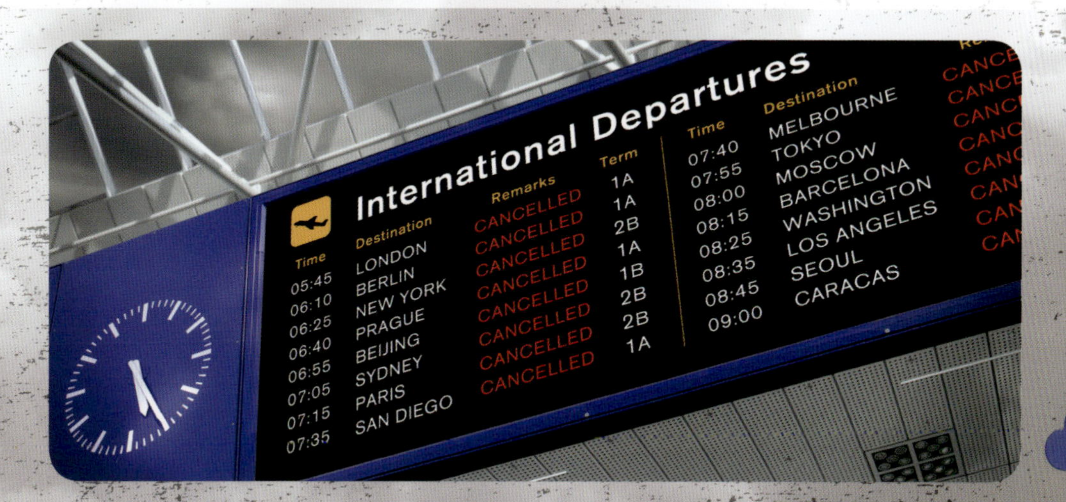

Thunderstorms can also be disastrous for businesses that operate outdoors, like farms and sports stadiums. The wind and hail from thunderstorms can destroy farmers' crops before they can be harvested, leaving them with less food to sell. Outdoor stadiums have to cancel or reschedule events like concerts and baseball games when there's a thunderstorm. They don't only lose the cost of customers' tickets but also the money that customers would've spent on parking, food, and merchandise like hats and T-shirts.

If a thunderstorm causes a power outage, grocery stores and restaurants may not be able to keep foods cold. The foods spoil and have to be thrown away, which wastes money that the businesses had already spent. They keep losing income as long as they can't sell anything to customers.

A storm at the wrong time or place can have a very negative impact on a business.

REDUCING THE IMPACT OF THUNDERSTORMS

Taking the correct steps can limit the damage that thunderstorms cause you, your family, and your community. Here's what you can do.

Severe Thunderstorm Risk Categories					
0	1	2	3	4	5
GENERAL THUNDERSTORMS	**MARGINAL RISK**	**SLIGHT RISK**	**ENHANCED RISK**	**MODERATE RISK**	**HIGH RISK**
VERY COMMON	COMMON	SOMEWHAT COMMON	SOMEWHAT COMMON	UNCOMMON	RARE
NO SEVERE THUNDERSTORMS No damaging or life-threatening storms expected	**A FEW STORMS COULD BE CLOSE TO SEVERE** No damaging or life-threatening storms expected	**SCATTERED SEVERE STORMS POSSIBLE** Significant damage or life-threatening storms unlikely	**NUMEROUS SEVERE STORMS POSSIBLE** Significant damage or life-threatening storms possible	**NUMEROUS SEVERE STORMS LIKELY** Significant damage or life-threatening storms possible	**WIDESPREAD SEVERE STORMS LIKELY** Significant damage or life-threatening storms likely
- Winds to 40 mph - Small hail	- Winds to 50 mph - Hail under 1" -Weakening storms	-1 or 2 tornadoes - A few reports of wind damage - Large hail >1"	- A few tornadoes - Several reports of wind damge - Large hail >2"	-Several tornadoes - Widespread wind damge - Large hail >2"	- Tornado outbreak - Derecho

Take watches and warnings seriously.

If a storm system is coming, make sure nothing of value is left outside.

If you have a well pump, store fresh water in case the power goes out.

Have a safe place to stay indoors during the storm, away from windows and doors.

Don't use anything that has to be plugged in, including landline phones.

If possible, listen to a cordless radio to stay updated on the storm.

Here's what to do if you are caught outside during a thunderstorm.

Seek shelter immediately. If you cannot go into a house, you will be safe in a car or truck as long as you don't touch any metal.

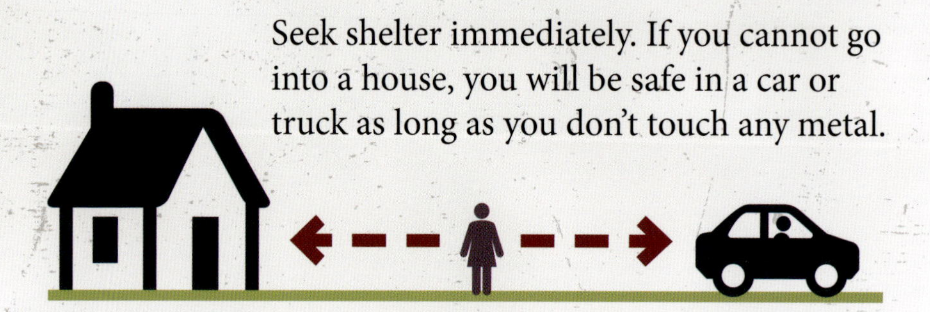

Avoid natural lightning rods such as trees and other tall objects.

If you are in the woods, find a low area where you can take shelter..

CONCLUSION

We may need thunderstorms to help provide fresh water for crops, but we have to be aware of their dangers, too. We need to listen to thunderstorm warnings and alerts so we can protect ourselves as much as possible. We need to have the right supplies ready to go before a storm comes. When we are safe, we can appreciate these natural wonders, instead of fearing them.

GLOSSARY

air mass: a large accumulation of air that is a different temperature or has a much different moisture content than the surrounding air

atmosphere: the layer of gases above the Earth that contains the air we breathe

derecho: a cluster of storms that can cover a large area and travel a long distance

haboob: a duststorm caused by downdrafts in a thunderstorm

jet stream: a fast current of air within the middle of the atmosphere

precipitation: any form of water that falls from a cloud

radar: a technology invented just before World War 2 that bounces a radio signal off objects to tell where and how far away they are

INDEX

RESOURCES

Challoner, Jack. *Hurricanes and Tornadoes.* New York: DK Books, 2014.

Davis, Barbara. *Air and Weather.* New York: Gareth Stevens Publishing, 2004.

Watts, Claire. *Natural Disasters.* New York: DK Books, 2015.

Williams, Scheley. *Tools of the Trade.* Logan, IA: Perfect Learning, Inc., 2004.

http://www.indiana.edu/weather.phenomena

http://www.tsgc.utexas.edu/thunderstorms

http://srh.noaa.gov/derecho.climo.html

http:// www.nssl.noaa.gov/research/thunderstorms/life.html

http://www.nssl.noaa.gov/research/thunderstorms/

http://www.insure.com/homeinsurance/costliestfire.html

https://ghrc.nsstc.nasa.gov/lightning/overview_lis_instrument.html

https://www.realclearscience.com/blog/2012/05/could-we-harness-lightning-as-an-energy-source.html

http://www.iowastateuni.edu/publichealth/stormprepare

NOTE: At the time of printing, all of the above resources were active and accessible. Due to the transient nature of the internet, we cannot guarantee they will be available in the future.